Animals that Build their Homes

A PRAIRIE DOG SITS OUTSIDE ITS HOME.

by Robert M. McClung

BOOKS FOR YOUNG EXPLORERS
NATIONAL GEOGRAPHIC SOCIETY

A mother beaver watches over her sleepy babies.
The babies are safe and warm inside their house.
Beavers build their homes in the water.
To build a house, the beaver cuts down trees
with its sharp teeth. Then it piles the wood in a pond.
Soon the pile is so big it sticks up out of the water.

A beaver swims out of its house.
The doorway is underwater.
But the beavers live
in the dry room above the water.

Some animals live in the open.
Some live in trees or caves.
And some animals, like the beaver,
build their own homes.

A young prairie dog slides down a tunnel into a tiny room.
It joins its brothers and sisters resting on a bed of grass.
A prairie dog makes its home by digging a burrow in the ground.
It carries soft grass down into its home and makes a cozy nest.

Other animals dig their homes in the ground, too.
A badger peeps out from the doorway of its underground home.
There is dirt on the badger's nose because it has been digging.
A chipmunk comes out of its burrow and looks around.
It lives in a nest in the ground all winter long.
When something scares it, down it goes. Soon it will be out of sight.

EASTERN CHIPMUNK

AMERICAN BADGER

The badger is a champion digger.
It can dig much faster
than a man with a shovel.
It uses its strong, sharp claws
and its pointed snout.

A crayfish sits all day in its home near a pond. It comes out of its hole at night and hunts for food. Soldier crabs dig their homes in the wet sand at the seashore. These two crabs are building round roofs over their burrows.

CRAYFISH

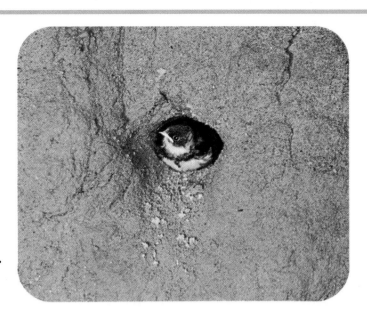

Some birds dig burrows.
A young bank swallow
peeks out of its home
in the side of a sandy cliff.

SOLDIER CRABS

The round hole
in the tree
is not too large
and not too small.
It is just big enough
so the woodpecker
can go in and out
of its nest
deep inside the tree.
The woodpecker
made the hole
by pecking
at the tree
with its pointed bill.
Its strong claws
help the bird
hold onto the tree.
The baby birds are
safely hidden
at the bottom
of the hole.

REDHEADED WOODPECKER

BOBOLINK

AMERICAN
REDSTART

12

ost birds build nests for their young.
Each kind of bird builds its own kind of nest.
The blue jay feeds its chicks in a nest of twigs high in a tree.
A bobolink makes a nest of grass on the ground.
A redstart uses strips of birch bark.

BLUE JAY

Six weaverbirds
hang by their claws
and flutter
their wings
outside their nests.
These nests
are open
near the bottom.
Many weaverbirds
build their nests
in the same tree.
The nests look like
fruit hanging
from the branches.
The birds weave
long strips of grass
and leaves
into a nest.

Flamingos use their bills to scoop up mud for their nests. The tops

of the nests are like cups, and the eggs do not roll off into the water.

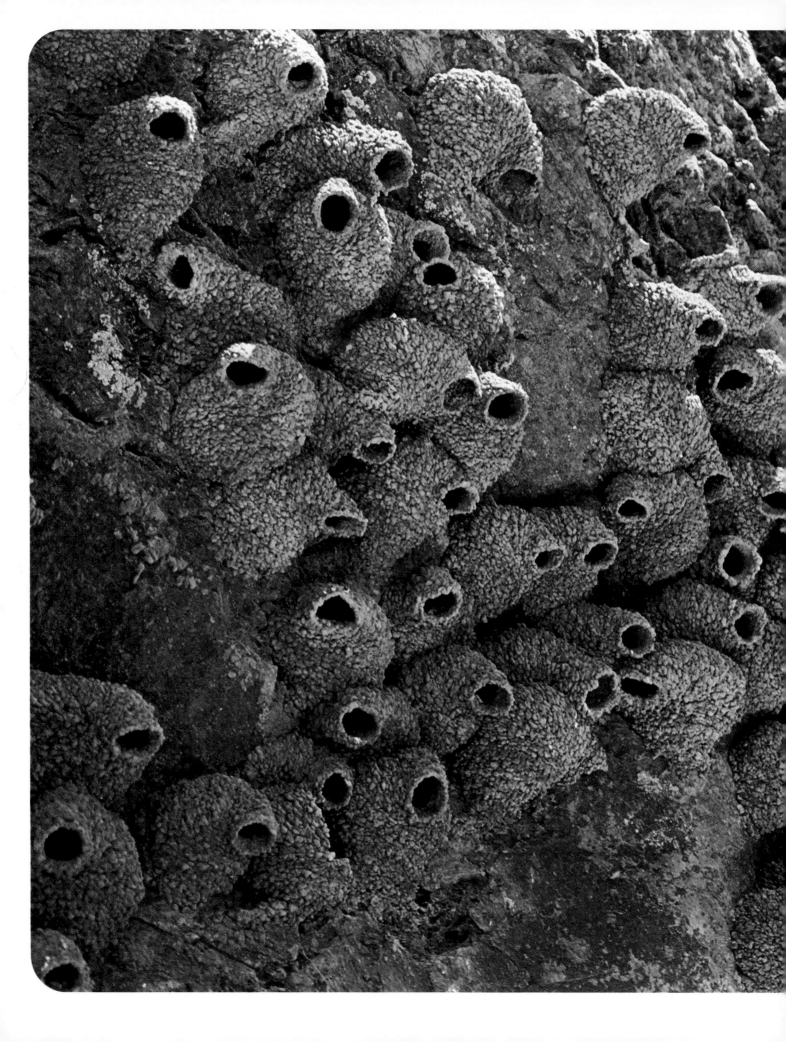

The mud nests of cliff swallows
seem to pop out from the sides of a cliff.
The swallows fill their beaks with mud
from a puddle or stream.
They use the mud to make their nests.
Bit by bit the nests are finished.
Do you see the bird peeking out of its home?

MUD DAUBER (ABOVE AND RIGHT)

Many wasps build nests of mud, too.

A wasp can work upside down. Can you?

The insect carries the mud in its jaws.

Soon the damp, soft mud will dry and become hard.

Some wasps build long tubes. Others make tiny round pots.

The mother wasp lays her eggs inside her nest.

NESTS OF POTTER WASP

NEST OF ORGAN-PIPE WASP

NEST OF MUD DAUBER

This hornet's nest looks like a big balloon.
To make a nest, the hornets chew wood
until it is a soft paste. They spit out the paste
and it dries into paper for the nest.
Inside the nest, there are many tiny cups.
Each cup holds an egg or a young hornet.

Wax honeycombs
hang from the ceiling
of a cave.
Honeybees make the wax
inside their bodies
and use it
to build the combs.
Each comb has
many tiny wax cups,
or cells.
Some cells hold honey
for food,
and some are homes
for baby bees.

Some termites build tall towers of earth.
A termite queen lives deep inside each tower.
Her long, fat body is full of eggs.
The tiny white termites have just hatched from eggs.
These termites work together to build their home.
They mix the soil with juices
from their mouths to make the earth sticky.

MOUND-BUILDING TERMITES

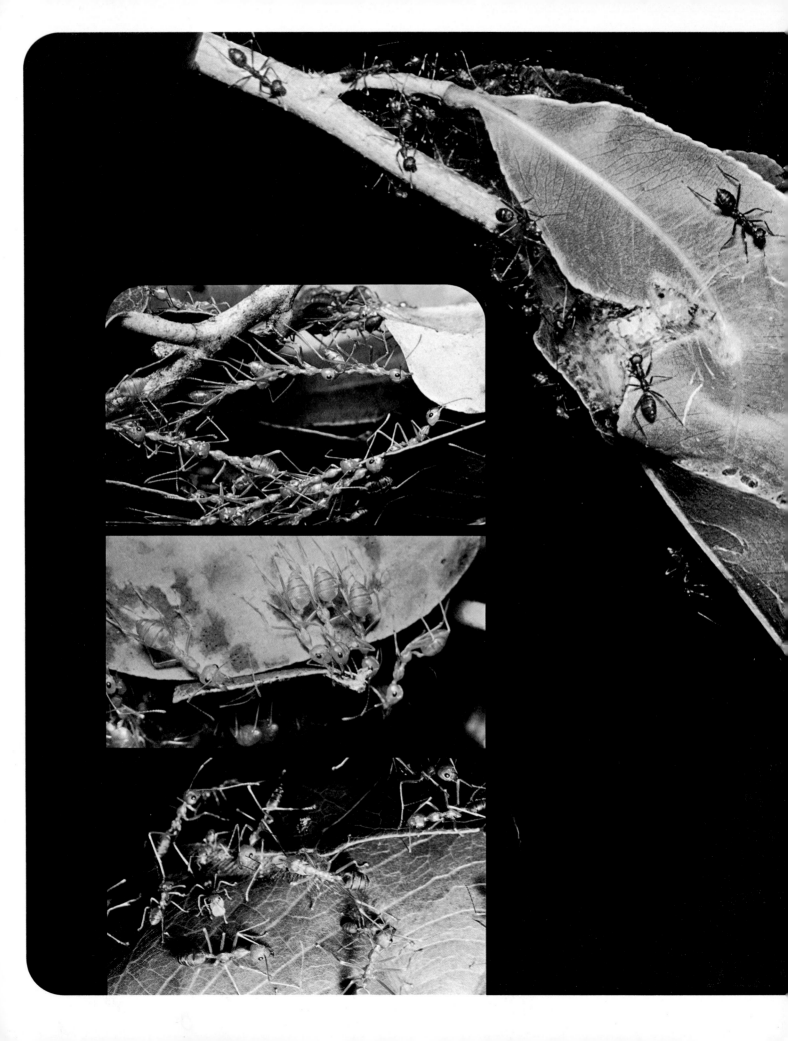

Weaver ants make their nests of leaves and white silk.
First the ants pull two leaves together.
Some workers hold young, white ants.
These young spin the silk that holds the leaves together.

A stickleback carries a weed
to make a nest.
The father stickleback squirts
a sticky glue from his body
to hold the nest together.
The mother fish goes into the nest
and lays her eggs there.
Sometimes the father spits sand
that covers the nest for a while.
Then he guards the eggs until they hatch.

Animals build their homes
without hammers, or saws, or nails.
They use different parts
of their bodies as tools.
Can you name some of the tools
animals use to make their homes?

THREE-SPINED STICKLEBACK

Published by The National Geographic Society
Robert E. Doyle, *President;* Melvin M. Payne, *Chairman of the Board;*
Gilbert M. Grosvenor, *Editor;* Melville Bell Grosvenor, *Editor-in-Chief*

Prepared by
The Special Publications Division
Robert L. Breeden, *Editor*
Donald J. Crump, *Associate Editor*
Philip B. Silcott, *Senior Editor*
Cynthia Russ Ramsay, *Managing Editor*
Toni Eugene, *Research*
Alice G. Rogers, *Communications Research Assistant*

Illustrations
Jim Abercrombie, *Picture Editor*
Josephine B. Bolt, *Art Director*
Suez Kehl, *Design Assistant*
Drayton Hawkins, *Design and Layout Assistant*

Production and Printing
Robert W. Messer, *Production Manager*
George V. White, *Assistant Production Manager*
Raja D. Murshed, June L. Graham, Christine A. Roberts, *Production Assistants*
John R. Metcalfe, *Engraving and Printing*
Jane H. Buxton, Stephanie S. Cooke, Mary C. Humphreys, Suzanne J. Jacobson,
Marilyn L. Wilbur, *Staff Assistants*

Consultants
Dr. Glenn O. Blough, Peter L. Munroe, *Educational Consultants*
Edith K. Chasnov, *Reading Consultant*

Illustrations Credits

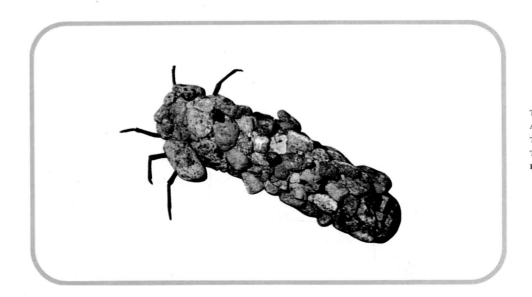

THIS CADDISWORM BUILDS
A CASE OF PEBBLES AND SILK.
THE SILK COMES FROM ITS BODY.
THE CADDISWORM HIDES
FROM ITS ENEMIES INSIDE ITS HOME.